Prayers For The Age Of Technology

Haskell M. Miller

CSS Publishing Company, Inc., Lima, Ohio

PRAYERS FOR THE AGE OF TECHNOLOGY

Copyright © 1998 by
CSS Publishing Company, Inc.
Lima, Ohio

All rights reserved. No part of this publication may be reproduced in any manner whatsoever without the prior permission of the publisher, except in the case of brief quotations embodied in critical articles and reviews. Inquiries should be addressed to: Permissions, CSS Publishing Company, Inc., P.O. Box 4503, Lima, Ohio 45802-4503.

Scripture quotations are from the *Revised Standard Version of the Bible*, copyright 1946, 1952 ©, 1971, 1973, by the Division of Christian Education of the National Council of the Churches of Christ in the USA. Used by permission.

Library of Congress Cataloging-in-Publication Data

Miller, Haskell M., 1910-
   Prayers for the age of technology / Haskell Miller.
     p.   cm.
   ISBN 0-7880-1172-3 (pbk.)
   1. Prayers. 2. Ecology—Religious aspects—Christianity. I. Title.
BV245.M52   1998
264'.13—dc21　　　　　　　　　　　　　　　　　　　　　　　　　　　　97-29583
　　　　　　　　　　　　　　　　　　　　　　　　　　　　　　　　　　　　CIP

This book is available in the following formats, listed by ISBN:
   0-7880-1172-3   Book
   0-7880-1173-1   IBM 3 1/2
   0-7880-1174-X   MAC
   0-7880-1175-8   Sermon Prep

*To my children
and their children,
with a prayer that they will be able
to live with poise, faith, and
hope in these changing times.*

Table Of Contents

Introduction	7
1. Dear Creator And Sustainer	11
2. Dear Wise And Purposeful One	12
3. Dear Mysterious Presence	13
4. Dear Caring God	15
5. Dear Patient Overseer	16
6. Dear Living Presence	17
7. Dear Supreme Valuer	18
8. Dear Master Craftsman And Innovator	20
9. Dear Lord Of Life	22
10. Dear God, By Whatever Name You Are Known	23
11. Dear Watchful And Caring Mother-Father God	24
12. Dear Loving, Merciful God	25
13. Dear Creator And Inhabiter Of The Universe	27
14. Dear Wonderful Provider	29
15. Dear Lord	30
16. Dear Concerned And Loving One	32
17. Dear Patient, Present, And Self-Revealing One	33
18. Dear Author Of Great Mysteries	34
19. Dear Love Entity	35
20. Dear Supreme Scientist Of The Universe	36
21. Dear Loving Communicator	37
22. Dear Father Of Us All	38

23.	Dear Ever-Present, Helping God	39
24.	Dear Merciful Judge And Loving Father	40
25.	Dear Source Of All Good Things	41
26.	Dear Compassionate Judge	42
27.	Dear Grantor Of Responsibility	43
28.	Dear Zealous Landlord	44
29.	Dear Benign Protector	45
30.	Dear Participant Observer	46
31.	Dear Source Of Truth And Reality	47
32.	Dear Demanding But Understanding One	50
33.	Dear God Of Gods	51
34.	Dear Almighty And All-Loving One	52
35.	Dear Infinitely Patient One	53
36.	Dear Master Communicator	54
37.	Dear Source Of Mystery	56
38.	An Easter Petition	57
39.	Science And Faith	58

Introduction

There have been times in the long course of human experience in which events have transpired that merit being labeled hinges of history. One was some ten thousand or more years ago, when the art of writing was developed, making possible the preserving of records of events in external form. Another was a few hundred years B. C., during the time of the Hebrew prophets, Buddha, and other religious leaders, when a great spiritual awakening occurred and ethical monotheism was identified. The life and influence of Jesus marked still another in the introduction of a powerful universalizing religion.

A development, so near our own time that we are still staggering from the shock waves produced by it, climaxed in what we have labeled the period of the Renaissance and the Enlightenment. This remarkable phenomenon, bursting forth only a few hundred years ago, accompanied development of the scientific method of gathering, checking, and verifying data. It revolutionized the process of reasoning, taking the guesswork and mystery out of a broad area of human experience, especially as related to the material aspects of reality. A philosophy of logical positivism was established, and the age of science was born.

The consequence has been nothing short of a great awakening. At long last humans had found a way of checking and verifying the data of experience concerning themselves and their environment. Many of the mysteries with which they had long been puzzled began to be clarified. Old superstitions, myths, fears, and bad judgments began to be dissipated. A fantastic age of innovation and discovery began. Unprecedented stimulation and freedom was experienced.

In particular, humans began to discover competencies they had not realized they possessed, and to assume responsibilities for which they had previously held deity or other supernatural influences accountable. Almost suddenly, they became more courageous and self-confident, more pleased with life on earth, and less anxious to escape to a life in the hereafter.

Human culture, which had been increasing at a relatively slow pace, took a giant leap forward, and it has continued to gain speed and momentum. Ways of thinking and acting were revolutionized.

To Dietrich Bonhoeffer, the brilliant young German theologian who paid with his life for defying Hitler, this development seemed evidence that humanity had "come of age." He saw it as a movement away from immaturity: from childlike reasoning and understanding to more adult-like informed and reasoned perceptions of reality.

Whether or not maturity has been achieved, there can be little doubt that the modern human stance, compared with what preceded it, is more like that of adults as compared with that of children. The so-called Age of Reason has been significantly different.

What will be the ultimate consequences of this remarkable change in human experience remains to be seen. Will human existence be enormously enriched, or devastatingly blighted? Logically, either outcome is, at this stage, a palpable possibility.

Even though it may not be possible to determine that it is genuine progress, progressive development unquestionably is occurring. The total mass of culture is being vastly expanded and enriched. Humans are much more at ease with their understanding of the physical universe. At the same time, they have lost many comforting feelings of certainty and security and are experiencing many new tensions and problems.

In any case, it seems apparent that, short of an eschatological termination, what the future for humanity will be is largely up to humans to determine. If, however, they can recognize and acknowledge that in their functioning and achievements they are not operating in a vacuum, but in interaction with the source and ground of their being, it will be significant evidence of their further maturity.

A major ongoing problem that has been greatly exacerbated is the conflict between reason and religion. It became much sharper as reason became informed and encouraged by the new science. Religious traditions, steeped in the thought forms of earlier generations, have not been easily reconciled to the new styles of thinking and dealing with reality. Making the problem worse is the fact that religion is concerned with the mystical and transcendental aspects of reality, which are not measurable by the methods of science.

Some persons, relying heavily on science and reason, are inclined to reject religion out of hand. Others are trying hard to reconcile the two. Their concern is to find the common ground shared by both, without compromising the integrity of either.

Such persons are anxious to maintain religious faith while being positively oriented in the ethos of science. The tensions they are experiencing make them acutely aware of the need of prayer. They want to be religious, but they want religion to make sense. They appreciate the methods and benefits of science, but they want science to be humble enough to recognize its limitations. They have concern with dimensions of reality that lie beyond the boundaries of science.

Above all, religiously concerned persons, caught in the tensions of this conflict, are aware of their need of help from the higher power that sustains and animates the universe.

There is challenge, however, in this sense of need. It is in recognition that the kind of prayer called for should manifest a quality of maturity befitting humans under the changed circumstances now existing. It should be sharing, conferring, advice-seeking communication on the part of mature persons who realize that they are being called on by the creative source of their being to take a large part in the process and responsibilities of creation.

Certainly, this type of prayer should be different from the childish, wheedling, would-be manipulative prayers so often characteristic of the past. They should be stalwart prayers of brave, courageous, committed people who are humbly anxious to be in partnership with the transcendent source of power on the frontiers of life.

Persons sensitive to the need of prayer in this mode will be prompted to petition, more earnestly than did the disciples of old: "Lord, teach us to pray!"

The prayers offered here represent a stumbling effort at praying appropriately in the changed circumstances that have been described. They struggle against the habits of long-established tradition and practice. At best, they are crude attempts to respond to sharply heard trumpet calls on today's real frontiers of human experience.

Walter Rauschenbush's *Prayers Of The Social Awakening*, written several decades ago, were in response to a major aspect of this same frontier. They were prayers in recognition of the more mature responsibilities Rauschenbush considered God was expecting of Christians and churches.

The attempt in the present group of prayers is to express something of the modified perspectives humans are acquiring on their relationship to the universe, and on the responsibilities entailed.

The writer's traditional Christian frame of reference will be recognized. An effort has been made, however, to keep in focus the broader human frame of reference, and the transcendental dimension of reality in the many forms in which it has been identified.

The awakening, after all, has affected the whole human family. It has been a human awakening, and its impact is just beginning to be felt by the whole.

It is the writer's sincere hope that these simple efforts at praying will be a useful stimulus to other Christians caught up in the same struggle he is experiencing.

1

Dear Creator And Sustainer:

By enabling us to develop the tools of modern science, you have expressed great confidence in us. You have let us in on many secrets of the natural order around us. You have blessed us with a more comfortable and satisfying physical existence. You have provided the means by which we can reason in a much sounder and more mature way. In short, you have made us feel like adults rather than children.

It is as though a concerned parent, anxiously watching his children grow, has concluded that they are sufficiently mature to be invited into limited partnership in the family enterprise. Happy and proud, the offspring, feeling pleased and honored, assume their place enthusiastically.

Limited experience and the immaturity of youth, however, are still very much with us. Our need of much patience and extensive supervision is very evident. Help us to control our brash, youthful impulses and to assume our newly assigned responsibilities in a commendable manner.

Guide us to a better understanding of how to relate to you in ways appropriate to our new level of maturity. Show us the best ways to use the new knowledge and skills we are acquiring. Give us the clarity of vision and strength of will to make right choices and decisions as we proceed.

Keep us reminded to try to be like Jesus in all that we do. Amen.

2

Dear Wise And Purposeful One:

Like sleepers suddenly aroused from dreaming, humanity is reacting with surprise and astonishment to this new dawn of awareness. People everywhere are taking stock anew of the world around them; they are celebrating freedom and opportunity they never knew before were available to them. They are busily engaged, also, in trying to shake off the burden of delusional dreams, myths, and speculations with which they have been comforting and beguiling themselves through the long ages of their history. Like children wide-eyed with wonder, they are re-assessing themselves and their world. They are convinced that they are being more realistic in what they see.

Your patience as you watch us plunge about in excitement is a sure sign of your love and grace. You understand us. You know that, as a whole, the human race is in an adolescent stage of development. You are not shocked that we are learning to say: "Oh, yeah? Who says so? And why?"

We trust your parental patience to understand that our attitudes and behavior are signs of increasing maturity. It is not so much evidence of doubt and defiance, we hope, as it is that we are growing up.

Help us understand our need of spiritual maturity to match our understanding of the physical aspects of reality. Bless us with a comparably enlightening spiritual awakening. Teach us to be humble as we experience growth.

In Jesus' name, Amen.

3

Dear Mysterious Presence:

As you surely have recognized, humans always have been aware of you, though your full identity has been a mystery to them. Early members of the race thought of you as a variety of spirits, present in rocks, trees, the wind, thunder, and almost every other physical phenomenon. In time, they gave the spirits symbolic human forms and organized them in a pantheon of deities. The prophets and other sensitive people perceived your unity and spoke of you as a single personal entity.

Jesus was recognized as your unique flesh and blood representation, but he made it clear that the one who sent him was much greater than he. Mystery, therefore, continues to surround your all-encompassing reality.

Like our ancestors of old, we sense your presence all around us. We think we have learned a little more about you than they knew, especially as your presence is manifest in your physical universe. But we realize that our understanding is far from complete. We are thankful, therefore, that you are continuing to reveal more of yourself to every generation.

The more we learn, however, the more we want to know. The mystery seems always to deepen. How great, indeed, you are! We look inside the atom and scan the vastness of the universe, finding your foot and finger prints everywhere. The recognition causes us to rejoice — and to wonder.

For a long while people have been comforted by thinking of you in very simplified terms. They have envisioned you as a person, with emotions and behavioral tendencies such as humans have. They have been happy to think of you as having supernatural power, and they have not hesitated to ask you to use it to change the course of events or set aside the order of the universe. They share a profound mysticism concerning you that is rooted more in their hopes and dreams than in substance-based faith.

Some people, however, having been aroused and challenged by the events of this modern era, are rejecting this traditional form of mysticism. To them, it does not have a rational basis. They spurn oversimplified myths. They dare to doubt, to ask questions, and to seek more logical answers.

These persons do not see you as one who can be persuaded to set aside the natural order of the universe. They do not spend the energy of their faith and wonder trying to flatter you or beg favors of you. They do not think of you as being so continuously angry that they must always be offering sacrifices to keep you appeased.

They believe the best way they can honor and please you is by making the greatest and most responsible use possible of the gifts of your grace. Using these gifts, they strive to gather and verify evidence on which to base a confident operational faith: a faith that will enable them to feel at ease in your presence and at home in your universe. Their desire is to be knowledgeable and mature children of yours.

While some become thoughtless and vain in this effort, as you know, most are serious and sincere. They live in the framework of the mystery, but are trusting you to help as they try to accept a full measure of responsibility for themselves and the welfare of the human community.

Rather than dreaming of escaping to ideal existence in a life to come, they strive for improvement of conditions in the here and now.

Stay especially close to them, Lord. They are at risk. They are pioneering on a new frontier, and they are in need of your Spirit's guidance. Help them to learn more about you as they learn about the world you made, and as they focus on the needs of the human community.

In the name of Jesus, who demonstrated concern for all our needs, Amen.

4

Dear Caring God:

We have awakened, but the dark mystery which still surrounds us causes us anxiety and fear. It is true that we have, in a sense, "come of age," but we are far from mature. Like venturesome adolescents, we pick our way uneasily, but with bravado, through dense jungles of persistent superstition, primitive misconceptions, stubborn dogmatism, and institutionalized error.

We stumble in uncertainty as we try to stand on our own feet. Exercising poor judgment, we make dreadful mistakes. While we are invigorated by a feeling of freedom, capability, and hope, we are greatly sobered by recognition of the burden of new responsibilities we have assumed.

Help us, we earnestly pray, not to let our immaturity breed recklessness that leads to disaster for humanity. Guide us through this adolescent stage of our development. Help us to achieve the maturity of a sincere commitment to the vision of the Kingdom of God on earth.

In Jesus' name, Amen.

5

Dear Patient Overseer:

In the chaos of change that surrounds us, it is comforting to believe that you are present and concerned; that you have a purpose and an ultimate end toward which you are working as you interact with us. Please do not let our careless exercise of newfound freedom or our juvenile obstreperousness distract you or divert you from that purpose.

We like to think we are not going against your will: that our efforts in physical science, technology, the science of human relationships, and in our avid quest for additional knowledge, are expressions of creative impulses flowing from you. As you will have noted, however, we are not doing too well with these privileges — these gifts of your grace.

Our efforts are unbalanced and erratic. They are producing much evil along with the good. Physical and material concerns tend to take precedence over the spiritual. Society's welfare and the welfare of persons is being subverted to material interests and the greed for gain.

Our desperate need is for a better quality of leadership: leaders with sufficient wisdom, sensitivity, and compassion to direct our new freedom and surge of creativity to more socially constructive and spiritually fulfilling ends. Dear perceptive and responsive God, help us produce such leaders.

Surround us with the persuasive power of your spirit, so that such persons may rise up among us: persons superior in leadership ability and spiritual sensitivity. And prepare us, as a people, to follow them in the direction of your vision for us.

Let there be a fresh awakening to the significance of your incarnation in Jesus.

In his name, Amen.

6

Dear Living Presence:

Save us from the hasty and superficial judgments some are making as we enter this new era of human experience. In the newfound freedom from primitive ways of viewing reality, some have lost sight completely of your unaltered presence.

Their expanding understanding of themselves and their environmental context has caused them to be afflicted with tunnel vision. In the resulting limited view, they see the universe as a mechanism, and humans as functioning in it in independence and solitude.

Disillusioned with naive traditional views, these sincere seekers of the treasures of truth have mistakenly concluded that you are dead, and that humans are alone and on their own in a purposeless universe.

Help these mistaken servants of yours to see that it is the oversimplified perception of you developed by our less sophisticated ancestors that appears to be dead. Relieve their agnosticism with better understanding. Correct the myopia which prevents them from seeing you clearly as you are, have always been, and will continue to be. Reward their earnest efforts with verifying data to undergird their understanding of aspects of reality that our ancestors struggled to interpret intuitively.

Help us all to recognize and acknowledge your living presence. Grant us to know with unequivocal certainty that you are with us and in us in all the experiences of our lives and in all the dynamics operating in your universe. Fill us with the joy of knowing that you are not an entity apart, but a living, present participant with us in the ongoing process of creation.

In the name of Jesus, who showed us your way, Amen.

7

Dear Supreme Valuer:

We have, in this new era, become aware of how our folkways, mores, values, and institutions were acquired. We now feel certain that they have been achieved, not given us pre-prescribed by you. This knowledge, however, has only added to our value problems.

When we were convinced that our values came directly from you, thinking about and dealing with them was relatively simple. You had defined and dictated them. All we had to do was decide whether or not we would respect and abide by them. Since they were backed by your authority, they were not to be questioned. They were absolutes.

This attitude has changed as human social experience has been subjected to scientific analysis. It has been recognized that humans have, from the beginning, had the decisive role in defining their values and establishing their institutions. Apparently, the valuing struggle has been an ever-changing, ongoing process. Obviously, you have not done the valuing for us, but have given us the ability to value and kept us relentlessly at the valuing task. Doubtless, our learning has resulted from the disciplining you have administered along the way.

Recognition of the human role in valuing, however, has made problems for us. We are uncertain and insecure, like children who have had their security blanket taken from them. We wonder if there are any value absolutes, or if all values are relative: to time, place, and the vagaries of human impulse.

Some persons have been tempted to rule out your influence, or even to deny your existence. Considering everything relative, they feel free to do more or less as they please.

Thank you, Our Creator, for giving us the freedom and responsibility of being our own valuers. Help us to be wise in their exercise. Give us the wit and wisdom to avoid moral chaos.

Help us to function free of the childishness of expecting you to do our valuing for us. Guide us through the jungles of relativism to mature understandings of reality. Guide us by your grace to discovery and celebration of what is supremely worthwhile, the things that have in them the quality of eternity.

In the name of Jesus, who helped us see so clearly the limitless value of love, Amen.

8

Dear Master Craftsman And Innovator:

Improvements of technology in the recent era have been so revolutionary that we are at a loss to know how to assess them. They have relieved us of much drudgery and vastly enriched our day-to-day existence. For all their benefits, we are profoundly grateful. Thank you for making the necessary channels of learning and growth available to us.

We are learning, however, that the benefits in technological change usually carry attendant risks and liabilities. In some of these negative consequences very serious problems are arising. Many are so great that we are beginning to wonder whether we have the ability to cope with them.

Primary among these are runaway population growth and pollution of the planet earth.

Improvements in food production and medical technology are improving health, saving lives, and increasing the life span, causing such rapid increase in population that it is being spoken of as an explosion. How long the earth's limited resources can sustain such growth is becoming a critical question.

The pollution being generated by the masses of people, and the technology necessary to sustain them, is already a major problem. Soil, air, and water, along with almost every other aspect of the environment, are being damaged at an alarming rate. Nuclear technology, in particular, has added enormously to the problem.

Thus far, humanity has found itself incapable of doing anything significant about these problems. We can see possible disaster ahead, and we recognize what might be done to avoid it. But we are so childish, so immature still, that we cannot reach consensus and take the corrective action called for.

With regard to the central issue, the population problem, we seem determined not to accept responsibility, and leave the matter up to you or some faceless working of fate.

Make us alive, dear Lord of Life, to the fact that our so-called coming of age has put us in charge even of this vital matter: the growth of our own numbers. Give us the maturity to accept this responsibility with proper regard for all the values involved. Let us not become indifferent to the sacredness of life. Give us the wisdom and courage to act beyond the inhibitions of tradition and apathy.

Prompt us with persistent reminders that we must accept this responsibility, and act upon it, or be prepared for the consequences of our sinful neglect.

Make us of the present generation especially aware of our obligation to help minimize the burden on generations to come. Amen.

9

Dear Lord Of Life:

The responsibility of our newfound freedom is frightening. Nowhere is it producing greater anxiety than in the fields of genetics and bioengineering.

While we are deeply grateful for the promise of relief from much pain and suffering, and for a more healthy and satisfying existence, the alternative possibilities are shocking to contemplate. It's great to think of being able to transfer genes and exchange body parts, and childless couples will rejoice to have test-tube babies. But we tremble with anxiety as we think of the possible consequences of our meddling with these basic elements of our creation and existence. Are we going too far? Are we intruding on ground that even angels might fear to tread?

We are very aware of the danger of creating causes of new diseases and unthinkable biological monstrosities. Already we see conflict developing over parentage of babies, and black markets in human body parts are operating.

Dear Holy and All-wise One, mercifully endow us with the maturity of spirit and judgment to use wisely and compassionately the knowledge and skills being developed in these fields. Keep us on guard against thoughtless and perverse ones who would use these abilities to evil ends.

Save us from the hell on earth that we could be creating.

In the name of all that is holy, Amen.

10

Dear God, By Whatever Name You Are Known:

Thank you for the challenge of problems to work on, and for the ability to cope with and solve them. They give zest and meaning to life. Thank you for sharing so much of your creative nature with us.

Help us to be humble and mature enough to use these gifts of your grace in a spirit of genuine gratitude. Help us to stay free of arrogance, and the foolish assumption that you have no part in the process. Strengthen us in faith and courage so that we may learn and grow, and function ever more wisely and creatively.

Keep us alert to the risks and dangers in our undertakings. Bear patiently with us, and forgive us, when we ignorantly or wilfully make less than the best choices.

Please know that we are both exalted and troubled by a feeling of being overwhelmed as we become increasingly aware of the many problems for which we are expected to assume responsibility. On the one hand, the opportunity they present is a welcome stimulus to creative functioning. But on the other hand, their complex enormity prompts us to wish for return to the simplistic state of our early ancestors, in which responsibility for such matters remained entirely in your hands.

Help us maintain mature faith and resolute hope. Steer us around the pessimism and despair of cynicism and hopelessness.

After all, was it not your purpose in putting us in the world to challenge us with problems, and experience with us the stimulus to creative response? Are we not expected to join with you in the fellowship of creativity?

We know we are in your hands, and we trust you to keep us in the paths of your purpose.

In the name of the one who showed us so plainly that you love us, Amen.

11

Dear Watchful And Caring Mother-Father God:
Like children maturing beyond the illusions of Santa Claus, we are growing away from perceptions of heaven and hell as literal places of reward and punishment. The realization makes us feel disoriented and insecure. How, now, should we think of the heavenly and hellish prospects of human life?

Since the dawn of value awareness, we have known that a process of judgment is at work in the universe. This fact seems to us one of the surest signs of your unwavering presence and concern. We are beginning to understand the concepts of heaven and hell as symbolic representations. They were adaptations to the level of limited human comprehension and maturity. As we become capable of looking beyond the symbols, however, to try to grasp the meaning of that which they represent, help us not to trivialize the concepts or minimize their significance.

Make and keep us ever more sharply aware that there are spiritual joys and rewards far more significant than gold and leisurely existence, and pain and suffering far greater than flames of fire can inflict.

Give us guidance for thinking and functioning more sensitively at the level of the spirit. Above all, keep us from wandering blindly or perversely into the "outer darkness" of alienation from you. Amen.

12

Dear Loving, Merciful God:

We are easily convinced that we are all sinners, for we are keenly aware that we are finite, fallible human beings. We compromise, equivocate, go against our better judgment, and often fail to abide by what we have reason to believe is your will for us.

But what exactly is sin? The ambiguity in the term troubles us greatly.

We are told that sin is "rebellion against God," but few of us have consciously and deliberately rebelled against you. Breaking the Ten Commandments is frequently cited, but strict legalism seems out of keeping with your expectation of obedient love, as set forth in the New Testament. We note that sin often is thought of as failure to abide by social customs, codes, or laws, but we know that customs and codes change from time to time and place to place; and man-made laws often fail to be fair or just.

The idea of original sin has become especially troubling. Is it a literal fact that sin is something born in us as a result of what Adam and Eve did in the Garden of Eden? We can see that there could have been no sin without understanding of the difference between right and wrong, good and evil. But this ability to understand seems more a blessing than a curse. Is it not evidence of your loving providence and grace that we have been endowed with this capacity? Is not this the point at which you have blessed us with the privilege of being your children, capable of responding to your love and enjoying meaningful fellowship with you?

How shall we reconcile the idea of original sin with the notion that we are personally responsible for sinning? Is not the Adam and Eve story an allegory with which early people kept themselves reminded of universal human weakness?

The evidence we are receiving through scientific studies in sociology, psychology, and biology, evidence we see as part of your continuing revelation, indicates that there are many factors affecting behavior that are beyond the individual's control. How shall these be taken into account? In fact, is not much sin social in nature, almost totally unrelated to individual responsibility?

It appears to us that the idea of original sin is related to the fact that we are born equipped with animal appetites and drives, and we must learn to function with conscience that struggles with values. Is not this the arena in which sin is defined?

We assume that you are always on the side of right and the highest values. It seems to us, therefore, that freely to choose less than the best in a situation where values are at stake would be the greatest kind of sin. It would surely be indicative of alienation from you and disregard of your will.

Bear with us, Lord, as you note our divergence from some of the views held by our ancestors. We have no desire to deny sin or to trivialize it. We seek a clearer understanding of it, and we are comforted by assurance that we can be saved from it. Amen.

13

Dear Creator And Inhabiter Of The Universe:

How shall we, in the light of our changing understanding of your universe and your reality, think of you and address you? Names suited to our old frames of reference seem not to fit anymore. We know that you are more than a person, that gender does not become you, and that you have not been adequately or accurately represented in the many other images we have used in the past.

We are wonderfully comforted and assured by the revelation you have given us in Jesus. He clarified for us the essential essence of your being, and made us aware of the potentials invested in us by your goodness and grace. We know him as your offspring: your expression of yourself in a form appropriate to our understanding. We are grateful for his opening to us a way into the intimacy of your family circle. It is joy to us to relate to you in Jesus' name.

Our need, however, is to know how to think of you and call your name in the light of our changing understanding of your vital presence in the circling stars, the smallest atom, and the heartbeat of living things.

Teach us how to address you and walk with you in this context. We celebrate with gratitude the fresh insights with which you are enlightening the mundane aspects of our lives. Help us with the etiquette that befits this new status.

Shall we continue to approach you in the traditional spirit of awe and obeisance? Or should we assume a more proud, erect stance, in awareness that we are your offspring, trying to accept your invitation to join you in understanding and fellowship? Are you actually an imperious and distant God, or better understood as a friendly and powerful presence?

Would you be pleased with such names, drawn from the language of everyday life, as: Maker, Creator, Friend, Provider, Nurturer, Almighty One, or the like? Would a name adapted from the technical terms of our science be more suitable? Or do you prefer that we continue with the pious terms in common usage?

Forgive us for beginning to struggle in discomfort with such names as "God," "Lord," "Father," "Holy Ghost," and others drawn from perspectives very different from those currently emerging.

In gratitude and faith, we seek your counsel and guidance. Thank you for your infinite patience and for giving us a place in your great cosmic plan. Amen.

14

Dear Wonderful Provider:

Thank you for arousing us from the sleepy ways of simpering childhood and challenging us to become strong, open-eyed, self-reliant adults. Thank you for helping us confirm so many of our age-old, intuitive insights and ideals, and for enabling us to free ourselves of so much handicapping ignorance, error, and superstition. Thank you for showing us how to understand our world and your universe better; for enabling us to free ourselves of great burdens of drudgery; for letting us learn how to produce material things in abundance; for giving us the means, and showing us the necessity, of living close together in a single community.

We know we have much more to learn, but these and many other attendant blessings seem enormously important to us.

In the excitement over these many benefits that have come so recently to us, however, many people have lost control or failed to achieve a needed sense of values. Yielding to temptation in the presence of unprecedented opportunities, they are selfishly grasping wealth and indulging their appetites. Materialism and hedonism, consequently, are threatening to overwhelm all other value concerns and override interest in things of the spirit.

Give us the willpower and strength to resist such temptations. Let your spirit work mightily in our hearts and minds until we gain a mature perspective on values and on our responsibility to one another in community. Amen.

15

Dear Lord:

Are we becoming irreligious because we are changing our perspective on human culture?

The view in the past, as we are keenly aware, was that the way of life practiced by a people was prescribed and directly given by you or some other mystical entity. Now, however, the prevailing view is that it is the product of group life. Our behavioral codes, and moral values, that formerly were considered commandments, laws, and values established by your decree, are now looked upon as folkways, mores, customs, and values developed in the course of our society's experience.

Does this mean a faithless denial of your role in human history? Are we to be condemned for it?

We can see the danger of a crude humanism concluding that there is no connection with anything beyond the limited horizon and time-bound plane of earthly existence. Save us, we pray, from such a spiritual catastrophe. Help us to discover and learn how to celebrate your presence and participation with us in ways not heretofore understood.

Forgive us for suspecting that we have been misled by past notions that the human world of culture must be considered evil — secular in contrast to sacred. We find it hard to believe that this is really condemned by you as "the world," from which we are commanded to "come out."

We recognize that error and sin make our shared way of life less than you desire it to be. But have you not been interacting with us through all the process of its development?

We think we sense your presence in: 1) the impulses and needs with which our original nature is endowed; 2) the drives that impel us toward experimentation, valuing, and fulfillment; 3) the capacity for discernment that makes choice and valuing possible; 4) the provision of an environment that lends itself encouragingly to our importunities; and 5) the gift of freedom to learn, and the ability to retain what is learned, through trial and error. You are, as we are beginning to see, always present with us, interacting with us encouragingly and judgmentally, in the culture-developing process.

Please do not look upon us only as sinners, obstinate and evil. Do not despair of us as we struggle to know you better, to understand better your continuing revelations of yourself.

Help us to know whether we are right in accepting these evolving views. Grant us relief from the anxiety we are experiencing as we yield to the compulsion to cast off moorings to contradictory traditional understandings.

In the name of Jesus, we pray, Amen.

16

Dear Concerned And Loving One:

Help us in our troubled understanding of Jesus. We doubt not that he was truly your offspring, all that a son should be. We admire and love him for the wonderful human being he was. But in what sense was he different from the rest of us? Was it an absolute difference, or only a difference in degree? We are aware of the record that tells us we too are your children and can become joint heirs with him in your favor and grace.

We understand him as human. But how shall we interpret his divinity? Was he divine in some mystical, supernatural way? Or shall we think of him as one who gave complete freedom to the divine in himself which exists as presence and potential in every person?

We know that Jesus loved us and gave himself for us. It was a wonderful act of saving grace. Nothing could have been a more perfect expression of the divine love that was in him. But was this a noble human act or a mystical, supernatural act, manifesting your purpose for our redemption?

We want to know Jesus better and to know you better through him. Please guide us in our quest.

We are so glad that he habitually identified himself as the "Son of Man." Was he not as truly a revelation of ourselves to us as he was of you?

In his name, Amen.

17

Dear Patient, Present, And Self-Revealing One:
Help us achieve a right perspective on the Bible. We respect and cherish it. We believe it is the greatest book ever written. But we are beginning to question traditional attitudes toward it.

Is it actually the Word of God? Or is it words about God? Is it literally true and infallible? Or is it a fallible human record interpreting human understandings of relationships with you?

More and more now, the Old Testament is being understood as the inspired record of a very sensitive and responsive people's historical struggle to know you and understand your will and way. That they received special insights is obvious: insights that serve as valuable guidelines and bases of hope for hosts of the world's people.

The many indications of human weakness and cultural bias contained in the record, however, are exceedingly troublesome. They raise questions about infallibility and about some of the claims concerning absolutes and supernatural experiences. Reason tells us that such claims were not dissimilar from common practice among early people who sought to explain experiences for which they had no better explanation.

Surely you are not as narrowly ethnocentric as the Hebrew people considered you to be, or as harshly exclusivistic as some of the New Testament record suggests. Such depictions seem very inconsistent with so much else that is understood concerning you.

Can we not celebrate the biblical record of revelation concerning your nature, character, and will without reifying and idolizing the book? Can not our faith be based in you, rather than in the book about you?

Help us, we earnestly pray, to be realistic about the Bible without losing faith in your living presence. Give us better understanding of how you have been continuously revealing yourself.

We make this plea in the name of Jesus, who, our faith tells us, was primary in your revelatory process. Amen.

18

Dear Author Of Great Mysteries:

A writer of a previous generation, trapped in the maw of racial prejudice, marveled that you would make a poet black, and bid him sing.

I marvel that you would choose a person with a thirst for knowledge, steep him in the ethos of modern science, fill him with expectations of objectivity and validation of data, and bid him witness to truths buried in the matrix of ancient miscalculations and mysticism.

It cannot be that you are a sadist who enjoys the sight of suffering. But what is the meaning? How should the fact of this agony be assessed?

Is it a test, like that which Job experienced? Is the hope that truth will be discovered, or faith refined, by the ordeal?

Is it possible that I, along with the many others like me, am being pressed toward a more mature level of faith and insight?

I am clinging, Lord, determined to hold on to faith. I trust you, though I cannot see you clearly, or understand you to the extent of my heart's desire.

Help me endure the pain and learn the purpose of it. Do not permit questions and doubts to become overwhelming; make them serve the cause of faith.

Keep me ever hopeful, and let me be an encouragement to fellow sufferers along the way.

A humbler, or more sincere, prayer I cannot offer. Amen.

19

Dear Love Entity:

Some who have known you best tell us that you are love. This comforts us, for, despite our would-be sophistication and skepticism concerning traditional concepts, we like to feel that a benign, caring presence animates the universe and looks after our lives.

We are, however, having trouble with the concept of love. There is an ambiguity in it that makes us uneasy. Our definitions leave us feeling that something is lacking.

We are familiar with the kind of love that draws men and women together romantically, the kind that exists among siblings and family members, and the kind that Jesus enjoined us to have toward one another as neighbors. But is there another dimension that will help us in our complex corporate life? This is where, in our present circumstances, the need is so great that it seems to mock all other expressions of love.

Can love exist in and between corporations? In international relations? Between races and ethnic groups? In a profit-motivated capitalist society?

Evidently, we are far from the full, mature understanding of love that we need. And our failure to implement it at any level is abysmally deficient.

Capable as we are in many respects, we seem incapable of functioning effectively at this primary level of need.

Help us to learn how to love: to be mature enough to be able to love as Jesus did. Give us the courage and will to know and embrace love as it exists in you. Amen.

20

Dear Supreme Scientist Of The Universe:
We thank you for sharing insights with those of us who are making this new age possible. It is so wonderful to be relieved of much of the drudgery, disease, and hunger that made life so difficult and hazardous in the past. Grant that as understanding of ourselves and your universe increases we may learn to know you and ourselves better, and walk more closely and intelligently with you.

Relieve the minds of those who are frightened by the strides humanity is taking. Many are feeling lost and insecure in the midst of the changes occurring. Give them comfort and assurance. Help them to understand that you are in the changes: the living, present, creator God of change.

So great are the changes, so unexplored the frontiers ahead of us, that all humanity is feeling exposed and at risk. Guilt feelings linger in the hearts of many, who suspect that we may be arrogantly defying our Maker. Yet, like children at play with new toys, we proceed with enthusiasm, assuming that we are making use of gifts of your grace.

Are not our efforts in line with the purpose you had in mind for us when you endowed us with the abilities being used? Is it not your will that we function with mature assurance, rather than with the uncertainty of infantile immaturity?

Be present with us, we pray. Teach us to stand tall and proud, in full awareness that we are your children.

We know that many of the things being done with science and technology are not of the highest order. Many are clear signs of evil in human hearts and in the structures of our social institutions. They, obviously, are not in keeping with your will. Help us to make the changes needed at these levels.

We are learning that we cannot expect you to handle for us the problems we are facing. In accepting the gift of freedom, we are becoming aware, we accepted the burden of responsibility. Give us strength for our tasks.

Keep us humble, but keep us proud to be your children. Amen.

21

Dear Loving Communicator:
We feel especially blessed by the changes which are occurring in transportation and communication. They have suddenly transformed our big world into a small neighborhood. New relationships and new styles of life are being developed; in some sense, they are being forced upon us.

We are fascinated by the promising potentials in what is occurring: better distribution of resources; better communication and increased understanding between peoples; more cooperation; and enrichment of life through sharing among newfound neighbors of treasured spiritual and cultural resources.

Many roadblocks, however, stand in the way of realization of these potentials. Obviously, we do not know how to live together in the new intimacy. Greed, prejudice, nationalism, and ethnic narrowness are among the prominent barriers. Transportation is being prostituted to greed and hoarding, rather than to uninhibited distribution. Communication, all too frequently, serves the purposes of propaganda, deception, and expressions of ill will.

Teach us how better to use these marvelous resources, how better to cooperate, and to express mutual respect and loving concern.

Motivate us to seek knowledge of you, your spirit and will, with the same energy and earnestness that characterize the scientific and technological quest. Amen.

22

Dear Father Of Us All:
Much to our surprise, we have learned that race is not what we thought it was. Our scientists are telling us that what we have been calling races are primarily conflict groups, with the conflict focused on some aspect of difference. Only sometimes is the difference a physical feature.

Careful research has shown that no reliable basis exists for identifying races according to physical differences. The physical features commonly associated with race you have caused to be widely diffused throughout the human population. They are, it has been discovered, actually distributed in a manner represented statistically by the bell-shaped curve of normal distribution. All human beings of all so-called races, we are assured, are far more alike than they are different.

Jesus and some of your more discerning servants of the past have called our attention to these truths. But most people have been unwilling or unprepared to believe and heed them.

Now a crisis is upon us. Human society is being devastated by racial and ethnic conflict, and prospects are that the situation could easily become much worse.

Increased mobility and modern means of communication have intensified intergroup contacts, and false assumptions about race have exacerbated old animosities. Powerful groups, exploiting others, have found in such assumptions convenient rationales for their actions, arguing that differences in social status are indicative of racially defined inherent difference.

Lord, have mercy on us. We are sick with the deadly virus of such attitudes. Help us as we struggle to rid ourselves of them and the many forms of discrimination and injustice which they spawn. Give us the will and show us the way to break out of the confines of racism, tribalism, nationalism, sectarianism, and all forms of narrow groupism in which prejudice is nourished. Amen.

23

Dear Ever-Present, Helping God:

Since the dawn of our era of so-called enlightenment, logical positivism has been an appealing option for many people. Humanism has described the universe as an impersonal mechanism with which humanity struggles alone, making its destiny whatever it can and will. Human minds and the universe, in this view, function according to physical laws and principles. No place is identified for you.

On the other hand, masses of people, puzzled and disturbed, cling to traditional views that you are present as a person, prepared to intervene actively in all that transpires. If they have misgivings about the logic in this faith, they do not acknowledge them. They persist, especially, in believing that you can be influenced by prayer to alter things in their lives and the world around them. While they are comforted by this belief, it poses many problems of logic for persons whose minds have become acclimated to the ethos of modern, science-oriented culture.

Some of us, Most Patient One, like persons seeing through a glass darkly, think we perceive you interacting with us, rather than intervening, in the established order. We sense that you are the dynamic in the order, participating cooperatively with us in the processes of life and change.

This is a tenuous view on which we desperately need further revelation. Help us achieve a fuller, more accurate, and mature understanding.

You, no doubt, are largely responsible for the fact that humanity is emerging from the wilderness of immaturity and misapprehension to this harsh frontier, on which the demand is for a more mature, clear-eyed faith.

Only your encouragement and assistance can prevent our being truly lost. Amen.

24

Dear Merciful Judge And Loving Father:

As you no doubt have noted, we humans, in this recent period of positivistic awareness, are deciding that our morals and values are of our own making and choosing. Always before, the assumption has been that they were givens, that you were to be credited or blamed for them. Conformity had resulted from unthinking, childlike obedience, or from fear of your supernatural power and judgment.

Now there is much uncertainty and disorientation. Many individuals are feeling released from supernatural sanctions and are running free. Their flaunting of long-standing moral codes and values is threatening the welfare of us all. Though you may not have been directly responsible for these traditional signposts and guidelines, they were the most inspired ones available. They had long been the foundation of stability and order in society. Thoughtless disregard of them poses the possibility of social chaos.

Help us to understand, and know how to deal with, this unprecedented change and the problems it is producing. Share with us at least a glimpse of your long-range view: what you are thinking, and working toward, in this disturbed and stressful era.

Are we humans becoming a hopeless basket case, going to hell on a rocket?

Or are we moving toward greater maturity, freedom, and assumption of responsibility? Are the present intimations of impending chaos merely evidence that we are in a period of transition to a new stage of development in your plan for us?

Dear God, grant that this may be so. Give us the will and faith to help make it so.

In the name of Jesus, Amen.

25

Dear Source Of All Good Things:

As you can see, Lord, science and technology have produced so many material benefits that, in the excitement, many people have lost control of themselves. Yielding to temptations of greed and selfishness, they are grabbing wealth and indulging their appetites. Others, experiencing the same temptations but being less successful in carrying them out, are feeling embittered and defeated.

In consequence, materialism and hedonism are running rampant in our midst. They threaten to smother concern for higher values and things of the spirit.

Send your Spirit among us. Let it work with the hearts and minds of us all. Help us to improve our sense of values, to be more concerned about the common welfare, to love one another more.

In the name of Jesus, who taught us the true meaning of things and love, Amen.

26

Dear Compassionate Judge:

Crime, we know, is nothing new to you. Since the episode between Cain and Abel it has been a persistent problem on the human scene.

In the present era, however, it is reaching epidemic proportions and threatening the foundations of organized community life. What should we do about it?

We make more laws, set up more courts, hire more policemen, build more prisons, and make punishment more severe, but crime continues unabated. Something, obviously, is amiss. What are we doing wrong or failing to do that would get at the root of the problem?

We recognize that in this new kind of technical society interdependence and inequity have been greatly increased. The masses of humanity are crowded together in a global village. Their value systems clash. Alienation grows where neighbors treat one another like strangers and have no common guidelines for the expression of hospitality. Resentment flares where deprivation exists in the presence of affluence and in awareness that the reasons for it are of human making.

In these and many other aspects of current reality, we are being made conscious of the fact that not all of the problem is evil in the hearts of the individuals who commit the crime. Much of the causation is in the deficiencies of the social system.

Help us, Lord, to learn how to control greed and callous exploitation of one another. Give us the will and the skill to eliminate discrimination and victimization, and the resentment and hostility that are generated by them.

Teach us to love one another more, so that all members of society will be more inclined to be cooperative and helpful. Amen.

27

Dear Grantor Of Responsibility:

In the assumption of a larger measure of responsibility for our world and our personal lives, we are launched on a series of new experiences. We have a new feeling of worthiness and self-confidence and are filled with enthusiasm for the life we have in this world.

At the same time, we are experiencing feelings of uncertainty and anxiety. It is harder to sense the security we enjoyed in our prior state of total dependence. We tend often to lose sight of you; we think we are alone and unsupported, totally dependent on our own strength. We wonder if we are causing you to be displeased with us.

Dear Lord, keep us headed in the right direction. Give us the benefit of corrective knowledge and a full measure of your forgiving grace.

Help us to gain confidence by learning to walk and talk with you as we explore the untrodden paths you open before us. Amen.

28

Dear Zealous Landlord:

Open our eyes of understanding to what we are doing to our environment: to the good earth you blessed us with the privilege of inhabiting. Shock us to awareness of the folly of polluting the air and water, depleting the soil and forests, wasting mineral resources, exterminating species, and the countless other thoughtless things we do. Shame us for the indifference we manifest for the welfare of generations which will follow us. Make us face the reality of what sorry, ungrateful tenants we are.

Please, however, do not summarily evict us, much as we may deserve it. Instead, give your blessings to those among us who struggle against great odds to make us change our careless, delinquent ways. And let the gift of your grace encourage development among us of leaders with the sensitivity and ability to inspire and help us become more caring and careful.

Help us to appreciate and enjoy this wonderful world, this Second Eden, which you have entrusted to us. Teach us the disciplines that will enable us to occupy it in a spirit of gratitude and good stewardship.

In appreciative love, Amen.

29

Dear Benign Protector:

Our forefathers lived in terror of wild beasts and nature's unpredictable whims. They trembled in awareness of their weakness and vulnerability. Understanding so little of the reality around them, they often considered you directly and intentionally responsible for what happened to them.

Now that our understanding of the natural world has improved, we are not so afraid of these simple dangers. We accept them as impersonal hazards and have gained much confidence in our ability to cope with them. Though we still need your help in dealing with the suffering caused by these so-called "natural evils," our better judgment tells us that they do not derive from specifically directed intention on your part.

We now live in terror at least equally great, however, from causes of clearly human origin. The sources are infinite, surrounding us on all sides and stalking us every day and night. We tremble in dread of such eventualities as horrible wars; nuclear disasters; crime that stalks our cities' streets; slaughter in the use of our transportation systems; hidden dangers in manufactured products on which we depend; exploitation by the powerful and greedy who stand poised to take advantage of our need and weakness; and so on and on.

You know, dear God, the awful things we are doing to ourselves — the monsters we create and tolerate in our midst. You know the anxiety in our hearts and the chaos and suffering in our communities.

We have not the audacity or presumption to ask you to solve these problems for us. But do help us acquire the wisdom, courage, and strength to deal with them and bring them under better control.

In the name of Jesus, who showed us how to proceed, Amen.

30

Dear Participant Observer:

If men of old made a terrible mistake in trying to reach up to you with the Tower of Babel, what of our undertakings in this modern day?

We have built towers many times higher; we have probed your universe in much more audacious ways. Where the men of Babel aspired to reach you, we have simply ignored you, as we have split the atom, learned to fly, put a man on the moon, and done countless other daring things in trying to ferret out the secrets embedded in your universe.

Will our efforts finally be dissipated in confusion, frustration, chaos, and despair? Or will we stumble in surprise upon your active presence and recognize your beauty and grace in it all?

Be with us, and bear with us, we pray, as we continue the exciting process of exploration and discovery. Bring us to awareness that we, too, like the men of Babel, are building a tower. Awaken us to the actual meaning in what we are doing: to the fact that we are striving toward you and an understanding of your purpose in creating us and the universe.

Do not count us arrogant or presumptuous. Help us learn the lessons of humility in recognition of our finitude. But give us the wisdom to know and celebrate the fact that we are your children, creative offspring of the Greatest Creator.

In the name of Jesus, our wonderful older brother, Amen.

31

Dear Source Of Truth And Reality:

By your grace, humans in this new age have been blessed with the discipline and tools of science, by means of which they have been enabled to do a better job of verifying their interpretations of their experiences. This means that they are now better equipped to think systematically and logically, to reason with more insight and maturity.

We, who are the beneficiaries of this development, are trying hard to develop the habit of looking for evidence that confirms and justifies views that we hold or hear expressed. We want them to make sense to us in terms of all data available. We are, at the level of our best thought and effort, concerned to make the facts of our experience and observation as intelligible as possible.

This, we trust, is in keeping with your will and purpose for us. We are comforted and encouraged by the apostle Paul's admonition to strive toward maturity: to cease being children, and to "grow up in every way" (Ephesians 4:12-15). He reminded us that it is important to have a reasoned faith that will enable us to develop toward "the fullness of Christ."

Yet we are troubled by the fact that there is conflict in our time over the impact that reason, supported by use of the methods of science, is having on traditional religious views and practices. Some seem to feel that the differences between scientifically informed reason and religion are irreconcilable, that there can never be a satisfactory accommodation between the two. Science, in their opinion, with its commitment to verifiable facts and objectivity, will never be able to reconcile itself to religion, which, as they see it, is based on myths and false or unverifiable assumptions.

Partisans on the side of religion are inclined to see science as a threat to most things long held sacred, especially to views that attempt to take the supernatural into account. On the other hand, partisans on the side of science are inclined to look upon religion as clinging to outmoded world views, and being an irrational, obtuse impediment to human progress.

Those of us who are devoted to religion and, at the same time, appreciative of science and grateful for its benefits, are confused and disturbed. We cannot believe that the differences actually are irreconcilable. Our conviction is that religion and scientifically informed reason share more common ground than is generally recognized. Grant us, we pray, the grace to discern that common ground more clearly. It seems the next step we must take toward greater maturity.

We recognize that religion's roots were put down in the experiences of early humans as they interacted with you in their environmental context. Because they were not sufficiently mature to interpret your revelations accurately, there were limitations in their understanding of their experiences. We can understand how some of these limitations became embedded in religious tradition. Though the traditions, therefore, may not be entirely accurate, they are at least authentic in that they represent sincere efforts to be properly related to reality.

Is this the same kind of effort in which we of this age, using the skills of science, are engaged? Are we not, like our ancestors, simply trying to improve our understanding of you and your creation?

Considerations such as the following encourage us to hope that reason and religion eventually can be made to function together effectively:
   a) the human need to know and be properly related to reality has remained the same through the ages;
   b) the reality context and you, its Creator, also have remained the same;
   c) modern efforts to explore and understand, though more sophisticated and accurate, are not, in essence, unlike the efforts of our ancestors among whom religion originated;
   d) so much of mystery in human experience of reality lies beyond the limits of scientific exploration that hypotheses comparable to those undergirding religion will continue to be essential to a reasonable understanding of the totality of reality.

Give us the wisdom to resolve this apparent conflict. Help us achieve the new level of maturity to which we are being directed by your Spirit.

In the name of the One who was the Word, the logic, that was made flesh to dwell among us, Amen.

32

Dear Demanding But Understanding One:

We think we understand that the reasons for crime are more complex than our ancestors assumed. We know that evil in the hearts of individuals is only part of the explanation. Evil in the structures and functioning of society also contributes to the problem. But we do not know how to make the needed corrections.

Obviously punishment does not deter, but we are committed to it as though it is our best, or only, option.

Help us to understand the impact of recent social change on the lives and behavior of individuals: the increased crowding and alienation; the technology that multiplies the means, opportunities, and temptations for exploitive behavior; the reasons for the breakdown in moral control and values; the dynamics involved in generating so much hostility in so many people.

We need no clearer reminder of our continuing immaturity than our failure in dealing with the crime problem. Let your spirit work mightily in our hearts and minds. Show us the way to move beyond competition, selfishness, and greed to an understanding of how society can, and should, be organized on a basis of love.

Help us become mature enough to acknowledge that blaming errant individuals so exclusively, and punishing them so callously, is a hypocritical oversimplification. Give us the will and wisdom to admit the broader basis of causation, and to work to correct it.

In the name of the compassionate Jesus, Amen.

33

Dear God Of Gods:
People from earliest times have sensed your existence and sought to interpret the evidence of your presence. Though their interpretations have varied greatly, each has been held with sincerity and reverence, though sometimes too with awe and fear. The resulting systems of religion have provided coherence and meaning to the societies in which they developed, giving evidence of your bountiful grace.

Have you not, in this universal process, been revealing yourself to all people everywhere, working with them to develop insight and understanding? Are you not pleased to note the earnestness with which all social units express their religious concerns, even though you may be seeking to improve the level of their insight?

And are you not also concerned, as we are being forced to be, that this universal religious earnestness tends toward idolatry of imperfect understandings? Does not the dogmatism and intolerance of the various religions displease you? Are you fully honored in such attitudes, no matter how sincerely they are expressed?

Dear God, above all our understandings of you, teach us tolerance of one another and appreciation of one another's efforts to be rightly related to you. Deliver us from the prejudice, hatred, and conflicts that are devastating the world of our time in the name of religion. Help us to understand one another and grow in the grace of humility and true ecumenicity.

In the name of the One you sent to show your love for everyone, Amen.

34

Dear Almighty And All-Loving One:

We need help with the concept of worship. It is such an ambiguous term.

If it means ascribing worth, there is little problem. We know your worth is beyond all valuing. Acknowledging and celebrating your worth, however, seems not to express adequately the relationship we enjoy with you: the relationship Jesus indicated for us, and of which we are increasingly aware as we appraise more realistically the evidence of your presence in your universe.

If worship implies subordination of self and submission to your power and authority, it would be appropriate, except for the fact that we have been instructed to think of you as a loving father. Is it appropriate for children to worship a parent while denigrating themselves and submitting unquestioningly to parental authority? Or is it more proper for them to accept with glad appreciation the care of loving parents, while growing toward the maturity of understanding and meaningful fellowship with them?

We know that worship as flattery or appeasement does not honor you or become us. We eschew it with disdain, while unconsciously drifting toward indulgence in it. Our primitive ancestors approached you in this way, but your continued revelations have suggested its inappropriateness.

Help us to find appropriate ways of expressing appreciation of your love, and of the privilege of fellowshiping with you in the enterprise of creation. Give us the increasing maturity that fosters understanding and sharing of purpose.

In the name of Jesus, your son and our loving brother, Amen.

35

Dear Infinitely Patient One:

Forgive us for what we have been, and what we still are, despite all your investment in us.

We marvel at your patience with us. You stood by us in the long ages of our primitivism and savage tribalism. When you thought we were mature enough to understand, you gave us Jesus to show us your hopes for us. Now you have entrusted to us the tools of science to enter many secrets of your creation; given us a larger measure of freedom and self-confidence; expanded the horizons of our reasoning.

How can you continue to trust us after the shambles we have made of the gifts of your grace? We killed Jesus. We have prostituted so many of the benefits of science to exploitation and injustice. Our reason has led us into dark forests of doubt.

Even we can see that, despite our haughty pride and veneer of civilization, we are still the same old savages we always have been. We are specializing in the creation of alienation, insecurity, and injustice. We are increasingly turning our faces from you to hedonism and secular concerns. We are saturating the earth with more and more blood spilled in violence.

Dear Source of Our Creation, and Sustainer of Our Continued Existence, do not lose patience with us. We know you have ample reason to do so, but we humbly pray for your continued mercy and grace. Let your boundless love continue to embrace us and teach us the lessons of love.

In the name of Jesus, Amen.

36

Dear Master Communicator:

Through the many wonderful ways in which you have maintained communication with us, our lives have been filled with assurance and joy. You have spoken through many sensitive souls in every generation, especially through Jesus, who was the Word made flesh and dwelt among us. We have ever felt the presence of your Spirit with us offering guidance and comfort. We have rejoiced in the privilege of interaction with you through prayer.

Your messages have come to us also through the many voices of nature, the rhythms of the heavens, and the orderliness observed by science.

Thank you for your comforting presence and for endowing us with the sensibilities that make it possible for us to hear and respond to you. Our desire is to hear you more clearly and respond to you with greater understanding.

Thank you, too, for making us social beings with the ability and need to communicate with one another and experience the joy of love. Thus you have redeemed us from dreadful loneliness and provided us the means for limitless enrichment and happiness.

We cannot thank you enough for the marvelous means of communication with which, by your grace, our creative efforts have been rewarded: the development of language; the art of writing; radio; television; satellite relays; cybernetics; the Internet; and proliferating other electronic means.

Recent development of massive, instant, worldwide means of communication has revolutionized the patterns of our lives. We are excited over the benefits being realized, and the potentials anticipated.

But, dear God, you can see better than we that all is not well in the uses being made of these media. Give us the wisdom and will to control their use for the common good. Help us avoid their prostitution to manipulative and exploitative ends.

And keep us alert to the danger represented in our great reliance on the machine in this arena of our experience. Guide us around the increasingly monstrous threat of letting machines so stand between us in our communication that they beguile us into the barrenness and numbness of machine-like interaction with one another.

In full awareness of our reliance on your guiding providence, Amen.

37

Dear Source Of Mystery:

You have revealed much to us, but the vastness of the still unknown remains an awesome challenge. It both stimulates us to continuous effort at new discoveries, and keeps us reminded of the limitations of our finitude. We thank you for making life so challenging and interesting. Your mysteries are an endless source of wonder and provocation to creative experimental effort.

Life itself is a mystery, and we pray for a right attitude toward it — an increased understanding of the meaning in it.

Death, too, is a problem. Despite Jesus' assurances, it remains an awesome mystery. Give us further help to conquer our fear of it. Enable us to perceive more clearly its meaning in the total complex of reality.

Reward our questing, we pray, with increasingly clear and reasonable bases for confident faith.

In the name of Jesus, Amen.

38

An Easter Petition

Come, risen Jesus! Rise up again!
And help us cope with the bloated monster of untimely death.
It strides in bloody arrogance across the land.
It nestles in the hearts of the greedy, who garner gain
    while masses go hungry.
It whispers in the ears of politicians that the rich
    should be pampered,
And the poor disregarded.
Its harvest is horrible in the slums of our cities,
On our roads, in our streets and airways,
And in the careless ways we daily live.
It pollutes the environment in gleeful abandon,
Excusing depredations with blind-sided logic.
It stalks through our factories and markets,
    grinning maliciously,
As we busily engage in administering prices and
    making munitions
To keep up employment and maximize profits.
It unceasingly sows seeds of hatred and malice,
Then glories in the stench that from our battlefields rises.
These terrors, we know, are of our own making.
We suffer because we have buried you so deep.
Please visit us again in a new resurrection.
Teach us once more how to love and have hope. Amen.

39

Science And Faith

Science stands forth in splendor, like a knight in
    full armor,
Slashing falsehoods and myths in the hard quest for truth.
Crowds in the bleachers, applauding the struggle, cheer the
    champ on, in hope of reward.
Strong knight of tradition offers mighty resistance.
Ignorance and error line up on his side.
Religion, in jealousy, being greatly offended,
Sponsors strong warriors to fight on each side.
Faith, the fair hostage, trembling and anxious, awaits
    the outcome.
Will the triumphing champion salute the fair lady,
Paying homage with favors beyond fondest dreams?
Or will he disdain her and leave her to languish,
Ignored and abandoned to live out her shame?
Lord, grant that this struggle be not fought in vain.
Bless the brave champion with a mind turned to wisdom.
Fill his heart with compassion for humanity's needs.
Wed him to faith in the triumph of your truth,
That all may be blessed by more knowledge of thee. Amen.

www.ingramcontent.com/pod-product-compliance
Lightning Source LLC
Chambersburg PA
CBHW071759040426
42446CB00012B/2623